YOUR KNOWLEDGE HAS VALUE

- We will publish your bachelor's and
  master's thesis, essays and papers

- Your own eBook and book -
  sold worldwide in all relevant shops

- Earn money with each sale

Upload your text at www.GRIN.com
and publish for free

Low Power Dissipation in VLSI Circuits. A Study of Low Power VLSI Design Techniques

Arpita Patel

Bibliographic information published by the German National Library:

The German National Library lists this publication in the National Bibliography; detailed bibliographic data are available on the Internet at http://dnb.dnb.de.

ISBN: 9783346949660
This book is also available as an ebook.

A Study of Low Power VLSI Design Techniques

Motivation

In the past, the major concerns of the VLSI designer were area, performance, cost and reliability; power consideration was mostly of only secondary importance. In recent years, however, this has begun to change and, increasingly, power is being given comparable weight to area and speed considerations. Several factors have contributed to this trend. Perhaps the primary driving factor has been the remarkable success and growth of the class of personal computing devices (portable desktops, audio- and video-based multimedia products) and wireless communications systems (personal digital assistants and personal communicators) which demand high-speed computation and complex functionality with low power consumption. In these applications, average power consumption is a critical design concern. The projected power budget for a battery-powered, A4 format, portable multimedia terminal, when implemented using off-the-shelf components not optimized for low-power operation, is about 40 W. With advanced Nickel-Metal-Hydride (secondary) battery technologies offering around 65 watt-hours/kilogram [52], this terminal would require an unacceptable 6 kilograms of batteries for 10 hours of operation between recharges. Even with new battery technologies such as rechargeable lithium ion or lithium polymer cells, it is anticipated that the expected battery lifetime will increase to about 90-110 watt-hours/kilogram over the next 5 years [52] which still leads to an unacceptable 3.6-4.4 kilograms of battery cells. In the absence of low-power design techniques then, current and future portable devices will suffer from either very short battery life or very heavy battery pack. There also exists a strong pressure for producers of high-end products to reduce their power consumption. Contemporary performance optimized microprocessors dissipate as much as 15-30 W at 100-200 MHz clock rates [20]! In the future, it can be extrapolated that a 10 cm2 microprocessor, clocked at 500 MHz (which is a not too aggressive estimate for the next decade) would consume about 300 W. The cost associated with packaging and cooling such devices is prohibitive. Since core power consumption must be dissipated through the packaging, increasingly expensive packaging and cooling strategies are required as chip power consumption increases. Consequently, there is a clear financial advantage to reducing the power consumed in high performance systems. In addition to cost, there is the issue of reliability. High power systems often run hot, and high temperature tends to exacerbate several silicon failure mechanisms. Every 10 °C increase in operating temperature roughly doubles a component's failure rate [63]. In this context, peak power (maximum possible power dissipation) is a critical design factor as it determines the thermal and electrical limits of designs, impacts the system cost, size and weight, dictates specific battery type, component and system packaging and heat sinks, and aggravates the resistive and inductive volage drop problems. It is therefore essential to have the peak power under control. Another crucial driving factor is that excessive power consumption is becoming the limiting factor in integrating more transistors on a single chip or on a multiple-chip module. Unless power consumption is dramatically reduced, the resulting heat will limit the feasible packing and performance of VLSI circuits and systems. From the environmental viewpoint, the smaller the power dissipation of electronic systems, the lower the heat pumped into the rooms, the lower the electricity consumed and hence the lower the impact on global environment, the less the office noise (e.g., due to elimination of a fan from the desktop), and the less stringent the environment/office power delivery or heat removal requirements. The motivations for reducing power consumption differ from application to application. In the class of micro-powered battery-operated, portable applications, such as cellular phones and personal digital assistants, the goal is to keep the battery lifetime and weight reasonable and the packaging cost low. Power levels below 1-2 W, for instance, enable the use of inexpensive plastic packages. For high performance, portable computers, such as laptop and notebook computers, the goal is to reduce the power dissipation of the electronics portion of the system to a point which is about half of the total power dissipation (including that of display and hard disk). Finally, for high performance, nonbattery operated systems, such as workstations, set-top computers and multimedia digital signal processors, the overall goal of power minimization is to reduce system cost (cooling, packaging and energy bill) while ensuring long-term device reliability. These different requirements impact how power optimization is addressed and how much the designer is willing to sacrifice in cost or performance to obtain lower power dissipation. The next question is to determine the objective function to minimize during low power design. The answer varies from one application domain to next. If extending the battery life is the only concern, then the energy (that is, the power-delay product) should be minimized. In this case the battery consumption is minimized even though an operation may take a very long time. On the other hand, if both the battery life and the circuit delay are important, then the energy-delay product must be minimized [26]. In this case one can alternatively minimize the energy/delay ratio (that is, the power) subject to a delay constraint. In most design scenarios, the circuit delay is set based on system-level considerations, and hence during circuit optimization, one minimizes power under user-specified timing constraints.

I.INTRODUCTION

low power has become a major subject in the electronics industries of today. For the design of VLSI chips, power dissipation has taken on equal importance to performance and area. The main issues below 90nm due to increased complexity are lowering power usage and overall power management on chip. Due to the requirement to lower package costs and increase battery life, power optimization is crucial for many systems. In low power VLSI designs, leakage current also has a significant impact on power management. An growing portion of integrated circuits' overall power dissipation is being accounted for by leakage current. This book discusses numerous power management techniques, methodologies, and tactics for low power circuits and systems. Future challenges for designing low power high performance circuits are also discussed. The benefit of combining low-power components with low-power design strategies is more important than ever before. As components get smaller, more battery-powered, and require more functionality, the need for lower power consumption is rising considerably. In the past, area, performance, and cost were the three main concerns for VLSI designers. The secondary problem was power consideration. Due to the extraordinary development and success of personal computers and wireless communication systems, which require high-speed computation and extensive functionality with minimal power consumption, power is now a major challenge. Applications have different reasons for wanting to cut back on power consumption. The objective is to maintain an appropriate battery lifetime, weight, and packaging cost for the category of micro- powered battery-operated portable applications, such as cell phones. The objective is to decrease the power dissipation ofthe electronics component of the system to a level that is around half of the overall power dissipation for high performance portable computers like laptops. The general purpose of power minimization for high performance non- battery operated systems, like workstations, is to lower the system cost while ensuring long-term device reliability. Process technology has pushed power to the forefront of all elements in such designs for such high performance systems. Power consumption from leakage has joined switching activity as the main power management concern at process nodes with technology below 90nm. Over the past ten years, a variety of strategies [15] have been created to address the continuously aggressive power reduction requirements of the majority of high performance. The fundamental methods of low power design, such as clock gating to cut down on dynamic power and multiple threshold voltage (multi-Vt) to cut down on Leakage current, are well known and supported by current tools [17].

2.STRATEGIES OF LOW POWER

Various methods for reducing power consumption are available at various stages of the VLSI design process shown in above Table I. Utilizing a variety of solutions at different stages of the VLSI Design process can result in effective power management. Therefore, designers must adopt a wise strategy for maximizing power consumptions in their creations.

Table I. Strategies for low power designs

Design Level	Strategies
Circuit/Logic level	Logic styles,transistor sizing and energy recovery
Operating System Level	Portioning, Power down
Technology Level	Threshold reduction,multi threshold devices
Software level	Regularity, locality, concurrency
Architecture level	Pipelining, Redundancy,Data encoding

3. POWER DISSIPATION IN CMOS:

There are two kinds of power dissipation in CMOS circuits: dynamic and static. Further dynamic power dissipation can be classified in two more categories: switching power dissipation and short-circuit power dissipation. The main source of switching power dissipation is charging and discharging of node capacitances (parasitic capacitances) of the circuit. Switching power dissipation component is represented by first term in the equation 1. Higher operating frequencies leads to frequent charging and discharging of node capacitances. It results in increased dynamic power dissipation. Hence to reduce switching power dissipation, switching activity of circuit nodes has to be reduced. The source of short-circuit power dissipation is short-circuit current flow from VDD to ground in CMOS circuits when both PMOS and NMOS transistors are simultaneously ON. The amount of short-circuit power dissipation depends on short-circuit current flow duration. Higher current duration leads to higher quantity of charge per transition, Qsc. Short-circuit power dissipation component is represented by second term in the equation 1. The source of static power dissipation is numerous leakage current flow in the OFF state transistor. The most dominant leakage currents are leakage current flow in the reversed biased PN junction and subthreshold leakage current. The total leakage current from various sources is denoted by Ileak in equation 1 and third term in this equation represents static power dissipation. We can sum up all above discussed power consumption components to get the total power consumption in CMOS circuit by following equation [1]:

$$P = \frac{1}{2}(C \cdot V_{DD}^2 \cdot f \cdot N) + (Q_{SC} \cdot V_{DD} \cdot f \cdot N) + I_{leak} \cdot V_{DD} \tag{1}$$

4. Transistor level techniques

4.1 Threshold voltage change

Higher as well as lower threshold voltage of MOS transistor both can benefit the reduction in the power dissipation. Higher threshold voltage transistors has lower subthreshold leakage current. This results in lower static power dissipation. On the other side, with lower threshold voltage transistors one can lower the supply voltage of the circuit which results in lower dynamic power dissipation.

4.2 SOI transistors

For many years we were using bulk transistor technology. Recently SOI (Silicon on Insulator) transistor technology has been developed. The main difference between bulk and SOI transistor is the presence of buried oxide layer below active silicon layer. Therefore, each transistor can be electrically isolated from the others [2]. Significant power reduction can be achieved by using SOI transistors because of lower parasitic capacitances and lower threshold voltage [2]. By using SOI transistors up to 50% active power can be reduced [3][4].

4.3 High-K materials for gate dielectric

In deep sub-micron technologies, gate dielectric thickness is very small just in the range of 2 to 3 atomic layers due to technology scaling. This results in significant gate leakage current if we are using low permittivity (K) insulators like SiO2. Most of the industries now replaced SiO2 by High-K insulators such as HfO2 and HfSiON to reduce the gate leakage even though these insulators has drawbacks like mobility degradation and increased short-channel effects.

5. Circuit level techniques

5.1 Transistor Sizing

Transistor size in a combinational circuit impacts the gate delay and the power dissipation. Wider transistors in a logic gate will have smaller gate delay, however switching power dissipation of the gate will increase. For a particular delay constraint, it is computationally difficult to find the size of the transistor that minimizes the power dissipation. One method to this problem is figuring the slack at each gate in the circuit. Positive slack provides us the margin by which gate can be slowed down so as its critical path delay is not affected. Circuits with positive slacks are processed, and the size of the transistors is reduced until the slack becomes zero, or unit the size of all transistors is smallest [1]. We can do transistor sizing for leakage power reduction also. Shorter channel length transistor will have more subthreshold leakage and thus leads to the more static power dissipation. However, on the other side, short channel transistors are faster because of higher saturation current. This leads to the trade-off between speed of the circuits and the static power dissipation [5].

5.2 Pin ordering

Figure. 1 shows NAND gate circuit designed using CMOS technology. A simple qualitative analysis of this circuit shows that for low power dissipation less charge transfer should happen in parasitic capacitances Cout and Ci . This can be achieved by applying high transition signal to the input A and low transition signal to the input B. This technique is known as pin ordering [5].

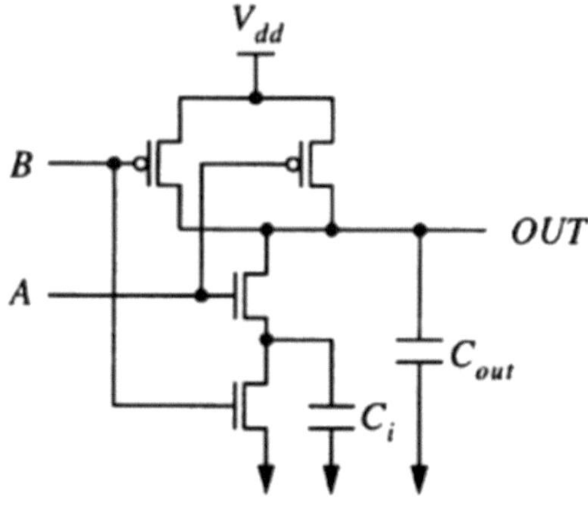

Figure 1: 2-Input CMOS NAND Gate[16]

5.3 Gate network reorganization

We can construct multiple gate-level circuits for the given function which are logically equivalent. Although these different implementations have same functionality, they may differ in delays and power consumption. Figure 2 shows two different gate-level networks with same functionality. It is obvious, network shown in Figure 2(a) will have more glitches than network shown in Figure 2(b). Thus network shown in Figure 2(b) will have less power dissipation compared to network shown in Figure 2(a).

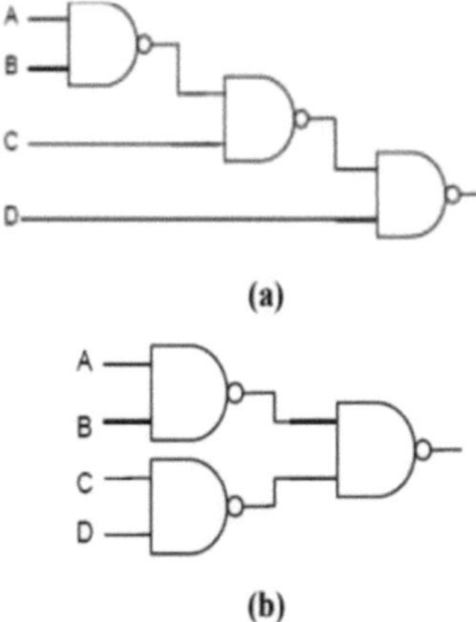

(a)

(b)

Figure 2:Two different gate level network for same function[16]

5.4 Multi-threshold (Vth) CMOS (MTCMOS)

This technique uses both low-Vth and high-Vth MOS transistors in the circuit. In the standby mode of the circuit both sleep transistors (MP and MN) are in OFF state to disconnect VDD and ground from the circuit. Both of these transistors are high-Vth transistors to minimize the leakage current. In the active mode of the circuit both sleep transistors are in ON state connecting VDD and ground to the circuits. Circuit can be implemented using low-Vt transistors to get the benefit of higher speed. Even in the circuit non-critical paths can have high-Vt transistors to minimize the power dissipation as delay is not the major concern on these paths [6][7]. One such example MTCMOS circuit is shown in the Figure 3 [6].

5.5 Transistor stacking

This is a technique in which two OFF transistors are stacked. This reduces the subthreshold leakage in active mode as compared to a single transistor. Two or more series connected OFF transistors will have less leakage current compared to single OFF transistor. There are two types of stacking: forced stacking and sleepy stacking. In forced stacking, a single transistor of width 'W' can be split into two transistors of width 'W/2' as shown in Figure 4 [8]. In sleepy stacking, first forced stacking is performed and then sleep transistor is connected in parallel

6

with one of the stacked transistor. This helps to reduce the propagation delay.

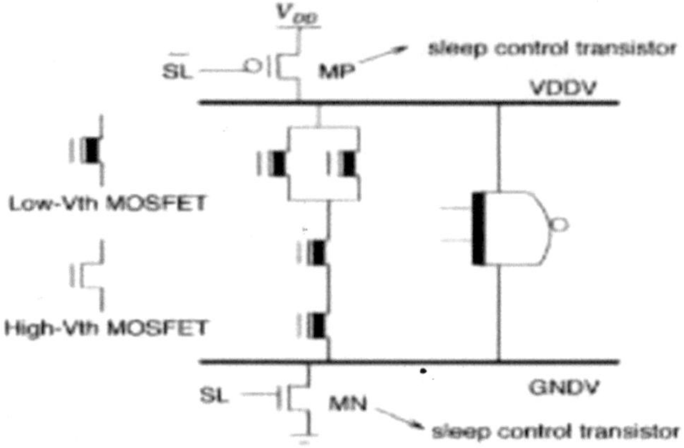

Figure 3-MTCMOS circuit[6]

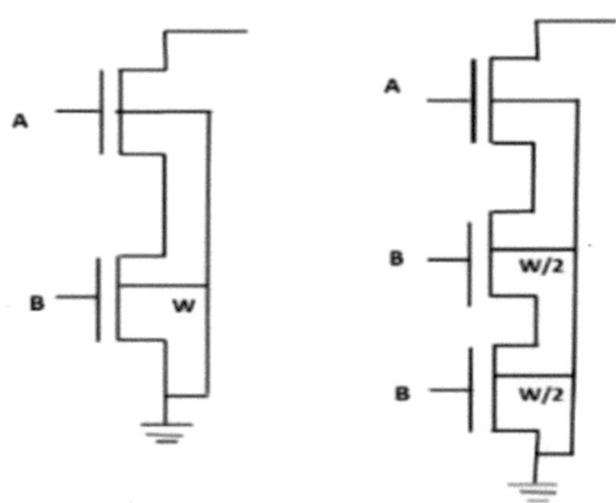

Figure 4:Forced transistor stacking[8]

6. Logic level techniques

6.1 State machine encoding

Two-bit binary counter will have states 00, 01, 10, 11, and again 00. Thus in four clock cycles there will be six switching activities. Now consider two-bit gray code counter which has 00, 01, 11, 10, and again 00 states. Here with gray code encoding, there are only four switching activities. Thus gray code encoding will have less switching activities and hence consume less switching power and more power efficient compared to binary encoding.

6.2 Bus invert encoding

It is a low power encoding technique first proposed in [9]. This technique is best suited to reduce the power consumption from the switching of data on off-chip buses. Figure 5 illustrates the architecture for this technique [5]. At each clock cycle, current and next bit at source end is observed and decision is made whether to invert next bit or not so that switching should not happen. If current and next bit are not same then next bit is inverted to avoid switching. A polarity status bit is send to the output side XOR gate to reinvert the inverted bits at source end. This technique is very efficient to reduce huge amount of dynamic power dissipation in case of transmission of high switching data on the buses.

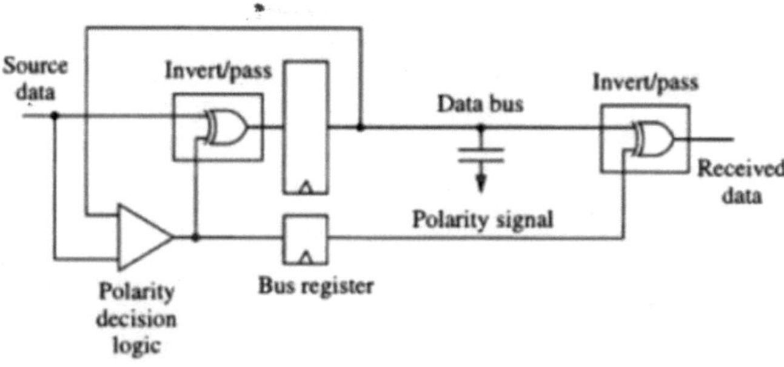

xFigure 5:Bus invert coding Architecutre[9]

6.3 Clock gating

One of the most efficient and popular clock signal power reduction technique is clock gating. When a clock signal to the functional block like memories, ALUs, co-processors etc. are not required for extended amount of time then we can mask the clock signal to these modules by gating. Normally NAND or NOR gate is used to stop the clock signal reaching to functional units. Figure 6 illustrates one such general clock gating scheme using NAND gate [5]. For many years clock gating technique has been widely used as dominant power reduction technique in processor design by researchers as well as by industries. Recently this technique is even used in the soft-core processor implementation on Artix-7 FPGA [10].

8

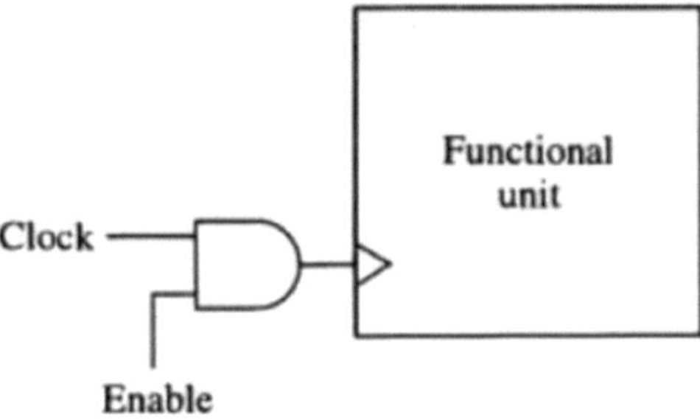

Figure 6: clock gating scheme [5]

7. Architecture level techniques

7.1 Pipelining for low power

Pipelining is normally used to increase the throughput of the system. But if there is no requirement of increase in the throughput, we can you pipelining to effectively reduce the power dissipation of the system. Two-stage pipeline architecture is depicted in Figure 7. We know that, the dynamic power dissipation of conventional CMOS circuit is given by

$$P_{conv} = C V_{DD}^2 f \qquad (2)$$

For an N-level pipelined system, its critical path is reduced to 1/N of its original length and the path capacitance is also reduced to (1/N)th of its original capacitance. If the same clock speed is maintained, then in the same amount of time, only (1/N)th capacitance has to be charged and discharged. This suggests that the supply voltage can be reduced to a new value (βVDD). Here β is in the range of 0 to 1. Hence the power dissipation of the pipelined architecture is given by [11]

$$P_{pip} = C \beta^2 V_{DD}^2 f = \beta^2 P_{conv} \qquad (3)$$

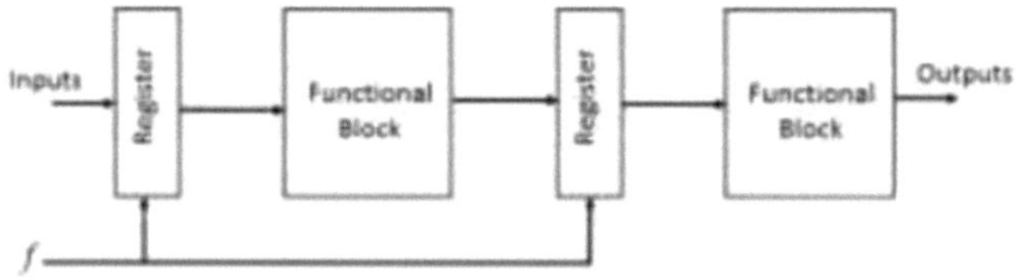

Figure 7:Two stage pipeline Amplifier[11]

7.2 Parallel architecture for low power

Parallel architecture, like pipelining can reduce the power dissipation by lowering the supply voltage. A two-stage parallel architecture is depicted in Figure 8. In an L-parallel architecture, operating frequency can be reduced to (f/L) without affecting the throughput of the system. As we can operate the system now at much lower frequency, now the VDD can be lowered to a new value βVDD because now to charge the same capacitance more time is available. Here $0 < \beta < 1$. Thus the power consumption of L-stage parallel architecture is given by [11].

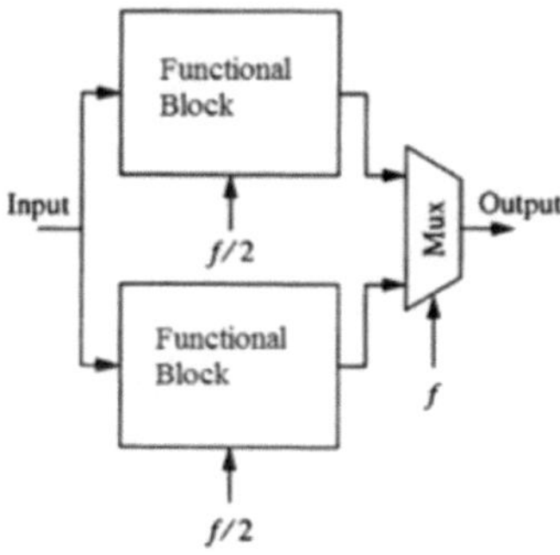

Figure 8:Two stage parallal Amplifier[11]

8. Advanced techniques

8.1 Adiabatic logic design

This logic design technique is known as adiabatic because it consumes zero power ideally. Static CMOS circuits are driven by a constant power supply. In this circuits there is significant amount of power loss as energy gets dissipated by channel resistance of MOSFETs during charging and discharging of load capacitance. To minimize this power dissipation energy-recovery adiabatic logic was originally proposed in [12][13]. Figure 9 shows a four-phase adiabatic inverter proposed in [12] and presented in [5].

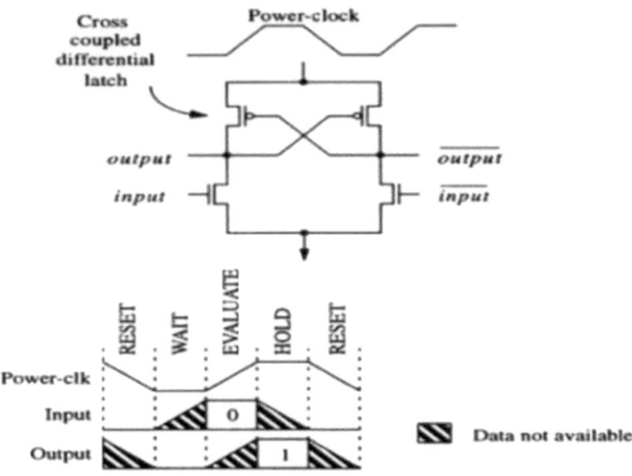

Figure 9: Four Phase Adiabatic Inverter [12]

A differential input is applied to the circuit and output is also differential. This circuit requires four-phase power clock for its working. Supply voltage is not constant. However a variable power-clock is used to power the circuit. PMOS transistors are cross-coupled with differential outputs to provide positive feedback. NMOS transistors acts as evaluation transistors during logic computation. Logic is computed during the EVALUATION and the HOLD phase as shown in Figure 9. The other phases are required for synchronous step-bystep operation of the circuit. The complexity of full adiabatic logic is high. The power-clock signal requires large number of phases for a moderately complex full adiabatic logic. A large amount of energy is wasted to generate many phases of power-clock signal. Hence quasi-adiabatic logic is more energy efficient at lower frequencies. A new technique to further reduce the power dissipation of quasi-adiabatic circuit is proposed in [14]. The power efficiency of adiabatic circuit is good only at lower frequencies. At higher frequencies energy consumption of adiabatic circuits became comparable to energy consumption of static CMOS circuits [5] as depicted by graphs in the Figure 10. This is the main drawback of adiabatic logic.

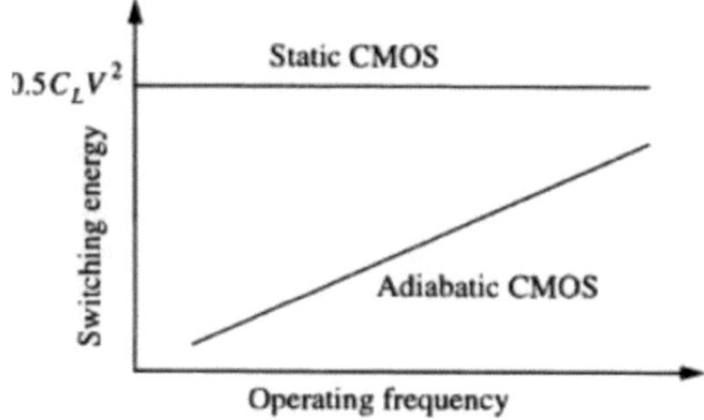

Figure 10: Switching Energy of Adiabatic logic and static CMOS [5]

8.2 Asynchronous computation

Today, a major portion of dynamic power dissipation in the chip arises from high frequency clock signal applied to synchronous processing unit. If we succeed to design asynchronous processing unit then there will be huge power saving as clock signal is not required. The speed of asynchronous computation unit is limited by the inherent delay of the circuit components [5]. A general block diagram of asynchronous computation unit is depicted in the Figure 11. The acknowledge and request signals synchronize the computation sequence and hence acts like a clock signal. Like clock signal there is no need of routing the acknowledge and request signals throughout the IC. Absence of high speed clock signal significantly reduces the switching activities in the asynchronous computation unit. This results in much lower power dissipation in the circuit. In [15] one such error correction code integrated circuit designed using fully asynchronous processing is presented. It reports, an 80 percent power saving when equated to the synchronous design.

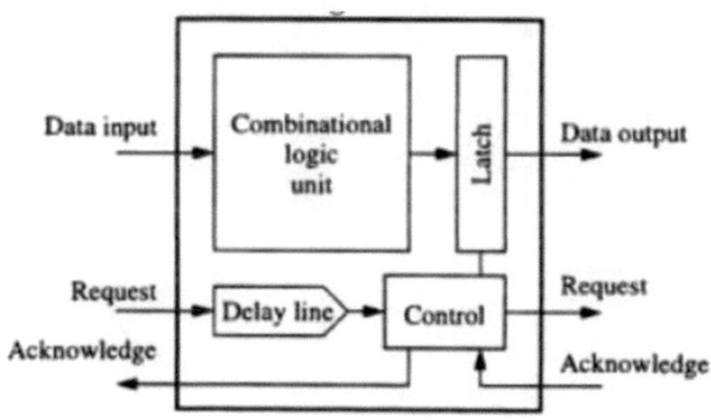

Figure 11: A General asynchronous computation unit[15].

9. Conclusion

Power optimization techniques at various levels of design abstraction is presented. Static power dissipation on the chip is increasing day by day due to very high level of integration. New manufacturing and design methods and techniques are evolving now a day to limit very high leakage current in the CMOS circuits. Today, high speed clock signal is the major contributor of dynamic power dissipation compared to data processing on the chip. Out of various power optimization techniques presented in this paper, the clock gating technique is commercially more often used technique. Adiabatic logic design and asynchronous computation are better choices but has many practical limitations. Software and operating system level power optimization techniques are gaining more attention now a day.

References:

[1] Srinivas Devadas and Sharad Malik, "A survey of optimization techniques targeting low power VLSI circuits", Proceedings of the 32nd Design Automation Conference, San Francisco, CA, (1995) pp. 242- 247.

[2] Christian Piguet, Marc Belleville, Olivier Faynot "Low-Power CMOS Circuits – Technology, Logic Design and CAD Tools", 1st Ed., CRC press, (2006). [

3] L. E. Thon et al., "250-600 MHz 12b digital filters in 0.8-0.25 μm bulk and SOI CMOS technologies", Proceedings of the Int. Symp. on Low-Power Electronics, (1996) August 12-14, pp. 89-92.

[4] M. Itoh et al., "Fully depleted SIMOX SOI process technology for low power digital and RF device", Proceedings of the 10th Int. Symp. Electrochemical Society, Washington, D.C., (2001) March 25-29, pp. 331-336.

[5] Gary K. Yeap, "Practical Low Power Digital VLSI Design", Springer, (1998).

[6] K. Roy and S. Prasad, "Low-Power CMOS VLSI Circuit Design", 1st Ed., Wiley-Interscience, (2000). [7] M. Anis, S. Areibi and M. Elmasry, "Design and optimization of multithreshold CMOS (MTCMOS) circuits", IEEE Transactions on Computer-Aided Design of Integrated Circuits and Systems, vol. 22, no. 10, (2003) pp. 1324-1342.

[8] M. Geetha Priya, K. Baskaran, D. Krishnaveni, "Leakage power reduction techniques in deep submicron technologies for VLSI applications", Procedia Engineering, Vol. 30, (2012) pp. 1163-1170.

[9] M. R. Stan and W. P. Burleson, "Bus-invert coding for low-power I/O", IEEE Transactions on Very Large Scale Integration (VLSI) Systems, vol. 3, no. 1, (1995) pp. 49-58.

[10] B. Tan, W. Lee, K. Mok and H. Goh, "Clock gating implementation on commercial Field Programmable Gate Array (FPGA)", Proceedings of the 4th International Conference on Electrical, Electronics and System Engineering (ICEESE), Kuala Lumpur, Malaysia, (2018) pp. 102-106.

[11] Keshab K. Parhi, "VLSI Digital Signal Processing Systems: Design and Implementation", WileyInterscience, 1st Ed., (1999).

[12] J. Denker, S. Avery, A. Dickinson, A. Kramer and T. Wik, "Adiabatic computing with the 2N-2N2D logic family", Proceedings of International Workshop on Low Power Design, (1994) pp. 183-187.

[13] A. Kramer, J. Denker, B. Flower and J. Moroney, "Second order adiabatic computation with 2N-2P and 2N-2N2P logic circuits", Proceedings of International Symposium on Low Power Design, (1995) pp. 191-196.

[14] Prasad D. Khandekar, Shaila Subbaraman, and Rajendra S. Talware, "Ultra-low power quasi-adiabatic inverter", Proceedings of Int. Conference on VLSI and Communication Engineering, (2009) April 16-18.

[15] K. Van Berkel, R. Burgess, J. L. W. Kessels, A. Peeters, M. Roncken and F. Schalij, "A fully asynchronous low-power error corrector for the DCC player", IEEE Journal of Solid-State Circuits, vol. 29, no. 12, (1994) pp. 1429-1439.

[16]Ketan J. Raut1*, Abhijit V. Chitre2, Minal S. Deshmukh3 and Kiran Maga," Low Power VLSI Design Techniques: A Review

YOUR KNOWLEDGE HAS VALUE

- We will publish your bachelor's and
  master's thesis, essays and papers

- Your own eBook and book -
  sold worldwide in all relevant shops

- Earn money with each sale

Upload your text at www.GRIN.com
and publish for free